# EARLY THEMES

# Homes

Ready-to-Go Activities, Games,
Literature Selections, Poetry, and Everything
You Need for a Complete Theme Unit

by Maria Fleming

SCHOLASTIC
PROFESSIONAL BOOKS

NEW YORK • TORONTO • LONDON • AUCKLAND • SYDNEY

Edited by Joan Novelli
Cover design by Vincent Ceci and Jaime Lucero
Cover art by Jo Lynn Alcorn
Interior design by Solutions by Design, Inc.
Interior illustration by Abby Carter

ISBN  0-590-99622-3

12 11 10 9 8 7 6 5 4 3 2 1      8 9/9/01/0

# Contents

# About this Book

What theme could be more central and relevant to children's lives than homes? More than physical structures, homes are family gathering places, and as such they account for a significant part of children's early experiences. Exploring this theme will allow you to connect studies with your students' world, providing rich and meaningful learning opportunities.

The early lessons in this book tap children's prior knowledge about the concept of home by having them take a close look at their own dwelling places. Students then venture out to explore homes around the world and in the animal kingdom, deepening their understanding of concepts as they investigate similarities and differences among shelters. As you teach the unit, allow students' curiosity to guide your study of homes. You can adapt the activities in this book as necessary to respond to their inquiries and learning styles, and to fulfill your curricular needs.

## WHAT'S INSIDE

The lessons, teaching ideas, and materials in this book include:

◎ ideas for setting up a learning center related to the theme of homes and activities to keep it going (see page 6) plus Learning Center Links throughout;

◎ an eye-catching poster to display, featuring Lucy Sprague Mitchell's poem "The House of the Mouse," plus related actvities (see center pullout and page 41);

◎ activities to launch your unit (see page 9);

◎ tips throughout for involving students' families;

◎ an essay on homes by Wendy Murray, plus two mini-books to make;

◎ Literature Connections throughout with related lessons to bolster reading skills;

◎ science experiments, math manipulatives, and social studies interactions as well as art, drama, music, and cooking activities;

◎ celebration suggestions for wrapping up the unit;

◎ age-appropriate reproducibles.

## WHY TEACH WITH THEMES?

Thematic teaching serves children's learning needs in a broad spectrum of ways. Building a unit of study around a high-interest topic—be it bugs or bears, homes or holidays—capitalizes on children's curiosity and motivates them to learn. In addition, thematic teaching allows children to revisit a topic over time and explore it through a variety of learning experiences, deepening their understanding of concepts.

The core of any theme unit is a wide selection of quality children's literature. Theme teaching enhances literacy development because it brings relevance to print. Children read, write, listen, and speak in a meaningful context as they acquire and exchange information about the theme and strive to answer their own questions.

Theme teaching also enables you to integrate the curriculum in meaningful ways as you plan activities and projects to investigate a topic in depth. This integrated approach helps children see connections among disciplines and to link learning with the real world.

But teaching with themes doesn't just benefit students—it's a boon to instructors as well. Teachers who use themes find they serve as a convenient framework for structuring lessons and organizing their day.

## GETTING READY TO TEACH HOMES

Following are some tips for preparing your unit on homes and suggestions for using the lessons in this book.

### Materials

Each lesson plan lists materials you'll need. Take a look around the classroom to see what you have on hand. You may also wish to send a note home to parents requesting contributions of materials. For example, several projects require boxes (shoe, cereal, gift, and so on), so it may be wise to start collecting materials several weeks before you begin the unit. Doctors' and dentists' offices may be willing to donate old copies of nature and decorating magazines, but again, this will require some advance planning. Some suggested materials are just that— suggestions. If they are unavailable, you can replace them with things you have on hand in your classroom or can readily obtain elsewhere.

### Grouping

You can approach the activities in a variety of ways. Many lend themselves to cooperative group or partner experiences, while others are best suited to independent work. For organizational ease, you may want to assign children to groups at the beginning of the unit. Grouping in advance will also prove helpful if you decide to set up a learning center on homes as part of your unit. (See page 6.)

## Assessment

The activities in this book provide many invitations for children to participate in discussions, respond to literature, create projects, explore topics in greater depth at a learning center, and more—all wonderful opportunities to assess learning. Here are two assessment tools you can use.

**KID-WATCHING CHART:** You might like to set up a simple assessment chart for kid-watching purposes throughout the unit. Here's one you can make. You'll need a 4-by-6-inch index card for each child and an 8-by-10-inch piece of oak tag.

◎ Tape the first index card to the top of the oak tag and label it Homes Theme Unit.

◎ Tape a second index card under the first, leaving about 1/4 inch of space between the cards. (See diagram, below.)

◎ Continue taping the cards to the oak tag in this way until you have a card for each student. Write each child's name at the bottom of a card, staggering the names as shown, below.

◎ Attach the chart to a clipboard and keep it on hand as you begin teaching the theme. You can jot notes on the chart as you move through the unit to assess each child's understanding of concepts and mastery of skills.

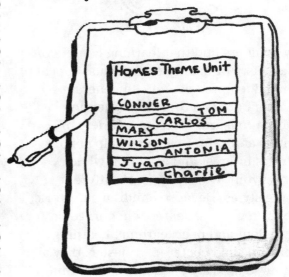

**INDIVIDUAL FOLDERS:** You may also want to have each child create a file for storing work as your class moves through the unit. Both empty file folders and large construction paper (folded and taped at the sides) work well. Have children write their names on their folders then decorate them with theme-related pictures. At the end of the unit, the folders will provide a convenient resource for assessing students' learning.

# SETTING UP A LEARNING CENTER

Many teachers find that learning centers offer big payoffs. For example, the visual appeal of a theme-decorated center can generate excitement about a topic and offer an invitation to learn. Offering a variety of activities at a center can also help capture children's interest and attention while allowing you to accommodate different learning styles. Because children can work alone or in groups, centers encourage both independent and peer learning. Learning centers are child-centered by design: they allow students to actively engage in the learning

process as you step back and observe.

Following are some suggestions for setting up and using a *Homes* learning center in your classroom:

◎ **SPACE:** Choose an area in your classroom that includes a work space (such as a large table or cluster of desks) and wall space or a bulletin board. You'll also want to provide chairs or floor mats. A bookshelf, on which to store and display books and activity materials, will come in handy. But cartons, baskets, or resealable plastic bags can serve this purpose just as well.

◎ **DECORATIONS:** Enlist children's help to create a bulletin board backdrop for the center that correlates with the theme. For example, you can create a house or apartment building shape using craft

## Teaching Tip

*As you plan your unit, you'll want to be sensitive to the fact that children live in different kinds of homes and come from different kinds of families. Students who have difficult housing situations may experience discomfort and ambivalence about the topic. Also, some children, sadly, may have no place to call home, or they may be shuffled among the homes of relatives and foster families. You can adapt the lessons in the book according to your students. Instead of having children build models of their own homes in the theme-launching project, for example, posture the activity as an exercise in imagination and have children construct dream homes. If there is a large problem with homelessness in your community, you may want to address this issue directly.* Fly Away Home, *by Eve Bunting (Clarion, 1991), handles the issue with sensitivity and makes a good springboard for discussing homelessness.*

paper. Use cardboard scraps to create shutters that open and close on the building. Children can lift the flaps to reveal people engaged in different household activities: eating dinner, watching TV, getting ready for bed, and so on. If there's room in your center, a furnished dollhouse will add a decorative element and invite interactive play.

◎ **READING MATERIALS:** Gather related children's books (fiction and nonfiction) and poetry and set up a mini-library at the center. (Literature Connections in each lesson will get you started.) Copy poems onto chart paper and hang them on the walls. The pullout poster features a poem you can add to your display: "The House of the Mouse" by Lucy Sprague Mitchell. Provide multiple copies of the Busy, Busy Beavers mini-book (see page 33) for independent, partner, or group reading.

◎ **ACTIVITIES:** You can use your center to introduce new concepts and to reinforce skills.

You may want to provide a single activity each day. Or, if space allows, you can set up a number of activities in the center and let children choose from among them. This book includes lots of reproducibles that are ideal for independent center work. In addition, several of the lessons lend themselves to learning centers, including Anybody Home? (page 28), Life in a Castle (page 30), and Nest Builders (page 38). A Learning Center Link at the end of each lesson plan offers more activity ideas. Shoe boxes or baskets make good containers for activity supplies. You may want to write the directions for each activity on paper and tape it to the shoe-box lid for easy reference. Walk the class through the center before students begin working on their own to be sure they understand how to use the materials.

◎ **SCHEDULING:** Set up and post a daily or weekly rotation schedule for center use.

Fill in the schedule with individual student's names, or group names, and times for use. It may be helpful to assign each group of students a color code, then color-code the schedule to provide a visual reminder of whose turn it is to use the center.

## Professional Resources

*Learning Centers: Getting Them Started, Keeping Them Going* by Michael F. Opitz (Scholastic Professional Books, 1994)

*Peterson Field Guides: Birds' Nests* by Hal H. Harrison (Houghton Mifflin, 1975)

*Is This a House for a Hermit Crab?* by Megan McDonald. Available as a Reading Rainbow video for $43.95. To order, call (800) 228-4630.

*The Multimedia Bird Book: An Interactive Field Guide* (Workman). A CD-ROM package that includes information on birds' nesting habits.

*Room to Live: Animal Homes and Territories—Mammals, Birds, and Other Animals* (National Geographic Society). A filmstrip, cassette, and teaching guide package.

**OTHER**
*Ant Farms*: available at World of Science stores, from Uncle Milton Industries, 3555 Hayden, Culver City, CA 90232; or NASCO, 901 Janesville Ave., Fort Atkinson, WI 53538.

# LAUNCHING THE THEME:
# There's No Place Like Home

Split-level houses, trailers, high-rise apartment buildings—the places we call home come in all manner of sizes and styles. But a home is more than a building that provides shelter. Day-to-day events—from sharing mealtimes together to a goodnight kiss at bedtime—shape our idea of home as much as the ceiling, floor, and walls. This section launches the theme unit with activities that invite children to take a close look at both the physical structures in which they live and the intangible qualities that define the word home.

## BACKGROUND NOTES

People build homes to protect themselves from the elements (rain, snow, wind, heat, cold) and intruders (other people, animals, and insects). Homes also provide a private place to rest, eat, and pursue leisure and other activities.

# Come Over to My House

**Students make mini-books about and models of their own homes, then create a display for sharing and comparing information about their dwelling places.**

## Materials

- mini-book (see pages 14–15)
- scissors
- stapler
- assorted boxes (shoe boxes, cereal boxes, sturdy gift boxes, and so on)
- glue, tape
- art materials (colored paper, paint, brushes, crayons, markers, and so on)

## Teaching the Lesson

**1** Before children start the project, begin a general discussion about homes: Why do people need homes? What kinds of buildings do we live in? Are people's homes all the same? Why might they be different?

**2** Help students make welcome mat-shaped mini-books to record information about their own home. Follow these directions to make the books:

- Make one double-sided photocopy of the mini-book for each child. (Be careful not to invert the text on the back of the page.)
- Place the side of the page with panel A and B faceup on the desk.
- Cut along the solid line to separate the panels.
- Place panel A on top of panel B.
- Fold the panels in along the dashed line to make the book.
- Staple along the left edge.

**3** Have children bring the mini-books home and enlist the help of a family member or caregiver to fill in the requested information. Children can draw pictures of, as well as write about, their homes and the people who live in them.

**4** Back in class, children can make mini-reproductions of their homes. Set up a craft table with the art materials listed. Invite children to choose boxes that best represent the shapes of their homes and use them to make their models. (For example, standing a shoe box on its end to make a high-rise apartment building.) Children can use assorted supplies to decorate their models, painting on details or using colored paper or cardboard to make roofs, windows, shutters, chimneys, and so on.

**5** Have students work together to arrange a display of the models, placing their welcome-mat booklets in front of their homes. You may want to make a banner announcing Wecome to Our Homes! to hang over the display.

**ACTIVITY Extension** Remind children that a home is much more than bricks and mortar (or wood and nails, or steel and cement). Refer children to the page of their mini-books on which they list the activities they do at home. Discuss how these activities help shape our idea of home. Provide each child with a brick-shaped piece of red construction paper (about 3-by-5 inches). Ask children to use a black marker or pen to complete this sentence: Home is where I _____.

Arrange the bricks in a house shape on the bulletin board. Strips of masking tape can serve as mortar to separate the bricks. Finish the display by hanging a Home, Sweet Home banner, designed by the children, over it.

**Literature Connection** Children are sure to enjoy the energetic verse in *A House Is a House for Me* by Mary Ann

Hoberman (Viking, 1978). It's also a great way to get kids thinking about the concept of shelter. After sharing the story, create sentence strips that follow this pattern: A _____ is a house for a (an) _____. Challenge students to add words that name homes for both people and animals. Place additional sentence strips at the learning center, along with cards on which you've written the students words. Invite children to complete the sentences on their own.

## Learning Center Link

*Create a sorting activity that spurs children's thinking about household necessities vs. household amenities. Together, cut out pictures of items such as plumbing fixtures, kitchen appliances, furniture, electronic devices (TV, stereo, and so on), toys, and books from decorating magazines. Paste each picture on a separate index card. Provide a surface on which children can sort the cards into items we need in a home and items we want in a home. Children can discuss their choices with partners.*

MATH

# House Count

**Children analyze data about their homes by creating and reading a pictograph.**

## Materials

◎ butcher paper
◎ tape
◎ sentence strips or scrap paper
◎ unlined index cards
◎ crayons or markers
◎ self-sticking notes

## Teaching the Lesson

1. To create the graphing surface, tape a large sheet of butcher paper to a wall (in a place students can reach) and divide it into five vertical columns, each wide enough to fit the index cards. Leave space along the bottom.

2. Write the following question on a sentence strip and hang it on the graph: What kind of home do you live in? (If students in your community all live in the same type of dwelling, graph a different question, such as What color is your home?) See Activity Extension, page 12, for other graph questions.

3. Label each of the columns at the base of the graph. Depending on the types of homes children live in, categories might include house, two-family house, apartment building, trailer, and so on.

4. Distribute one index card to each student. Ask children to draw their homes.

5. Invite children to gather around the graph with their index cards, then take turns taping the cards in the appropriate columns.

6. Ask volunteers to tally the number of index cards in each column. Use self-sticking notes at the top of each column to indicate the totals. Guide children in reading the graph by asking questions such as:

◎ How many people live in apartment buildings?

◎ Do more people live in one- or two-family houses?

◎ Do fewer people live in apartment buildings or houses?

◎ How many fewer people live in trailers than apartment buildings?

**ACTIVITY Extension** Graph other information about students' homes. Instead of pictures, use symbols (such as stickers or checkmarks) or tally marks to indicate the information on the graph. Invite children's input in choosing graph questions. Some possibilities are:

◎ What is your home made of? (wood, brick, concrete, and so on; children can ask a family member if they are unsure).

◎ What color is your home (inside/outside)?

◎ How many people live in your home?

**Literature Connection** *Anno's Counting House* by Mitsumasa Anno (Putnam, 1982) gives students opportunities for counting, adding, and comparing groups of people as they move from one house to another. Ask children to write addition and subtraction sentences for each page of the book. Provide counters to support their work.

## Learning Center Link

*For another math-related activity focusing on the theme of homes, provide pattern blocks that children can use to configure different house shapes.*

# A Sense of Place

**Students listen as you read "From My Room" by Wendy Murray, and try to identify the sounds that the writer associates with her childhood home. Children then describe sensory experiences that make their own homes unique.**

## Materials

◎ "From My Room" read-aloud (see page 16)

◎ Sounds Like Home reproducible (see page 17)

## Teaching the Lesson

1 Try a visualization exercise to get students to focus on the sounds they associate with their homes: Ask children to close their eyes and imagine themselves waking up in the morning. What sounds do they hear as they lie in bed? What sounds do they hear as the people in their home get ready to begin the day? (muffled voices, pans clanging, water whooshing, and so on). Tell children you are going to read them an essay by author Wendy Murray. In it, the writer talks about the sounds she remembers from her childhood home.

2   You may want to give children a sense of place before reading the essay by locating Connecticut on a map. Can children find the Connecticut River?

3   Read aloud the essay. Ask children to pay close attention to the sounds the author describes. Can they remember words for some of the sounds? (whirring, thunder, shake, clanking, creak, and so on) What are some other ways the author describes sounds? (She talks about the sounds of breathing, the crickets and cicadas, the sound of her mother walking across the deck, and so on.)

4   Read the essay again. Can students remember which sounds the author heard in the morning? In the evening? Other times?

5   Distribute the Sounds Like Home reproducible to children. Ask them to imagine they are home on a Saturday or Sunday. What sounds do they hear at different times of the day? Children can record these sounds on the paper. If children have trouble remembering sounds, you may want to have them make pocket diaries to take home and record sounds in. They can add these sounds to their reproducible charts in class. Be sure to allow time for children to share the sounds they record.

**ACTIVITY Extension** Sounds are just one of the ways our homes speak to us. Ask children what smells remind them of their homes. For example, is something often baking in the oven when they get home from school? Does a restaurant fill their home with the aromas of pizza or other foods? Can they recall the smell of a special soap in the bathroom, of scented candles scattered around they house, or the way the towels smell right out of the clothes dryer?

**Literature Connection** Share selections from the many wonderful essays, poems, and illustrations by well-known children's authors and illustrators in *Home* (HarperCollins, 1992). Several poems and essays describe an author's favorite spot in his or her home, be it the attic, a closet, or a comfortable chair. After reading these pieces aloud to children, invite them to describe—using art and or words—a favorite place in their homes.

Here I am at home! (draw a picture)

7

Some of the things we do in our home are:

Welcome to

_____
(your name)

's home!

2

The people who live here are:

5

I live in

_____

(type of home)

We have lived here for

_____

(number of years, months, or days)

1

3

The thing I like best about my home is

_____

_____

_____

_____

Our home has _____ rooms.

(number)

# From My Room

by Wendy Murray

My bedroom was on the second floor of a house overlooking a backyard and the woods. A half mile away and hidden by trees was the Connecticut River. My three brothers and I each had bedrooms along a small hallway. On the hottest summer nights, my father would place fans in the doorways of our rooms and we would bring our pillows down to the floor and fall asleep in the doorways, drifting off to the sound of the whirring fans and one another's breathing.

My brothers were not always sleeping angels, though. They would roughhouse in the hallway. The walls of my bedroom would thunder and shake. I'd shout for them to stop, dancing around them like a boxing referee.

From my bedroom I could hear our dog Nanny eat her dinner, her dog tags clanking against the metal bowl. I could hear crickets and cicadas in the summer and the creak of trees on cold winter nights.

I had a white clock radio beside my bed. I loved to listen to the news and music. But even top 40 hits weren't enough to rouse me from my bed on school mornings, especially when it was cold outside. My father had the ultimate alarm clock, though. After he came back from jogging, he would throw a basketball against the side of the house by my window. The glass would shake, the house would quake, and I'd jolt awake and get ready for school.

# Sounds Like Home

What sounds make you think of home? Are there special sounds you hear at different times of the day? Write your ideas in the boxes. You can draw pictures of the sounds too.

**Morning Sounds**

**Afternoon Sounds**

**Evening Sounds**

**Anytime Sounds**

# HERE, THERE, AND EVERYWHERE:

# Homes Around the World

From the mud huts of the Dabdam of Ghana to the goat-hair tents of the Bedouins of the Middle East, homes around the world are as different as the people who live in them. Activities in this section will help your students compare and classify the different structures people around the world inhabit. Your students will also investigate building materials and consider different cultural aesthetics that dictate home design.

## BACKGROUND NOTES

The types of houses people build are influenced by a number of factors, including materials available, physical environment, and climate and weather conditions. Shelters made from palm leaves and grasses that are found in tropical climates, for example, would not be suited to arctic environments. Likewise, traditional Inuit igloos would be impractical—as well as impossible to make—in balmy climes.

# The Wide World of Homes

**Children create an interactive bulletin board display of homes around the world, which they use for sorting and classification activities.**

## Materials

◎ magazines with international themes (such as *Faces*, *National Geographic*, *National Geographic World*, and *Traveler*)

◎ travel brochures (ask a local travel agent to donate materials)

◎ scissors

◎ unlined index cards

◎ glue

◎ hole punch

◎ string

◎ pushpins

## Teaching the Lesson

1. Hunt through magazines and brochures for pictures of homes in other countries. Try to find pictures of homes from a wide cross-section of cultures and in a wide variety of styles, shapes, and sizes. Cut out the pictures, then paste each one on an index card and label it with the name of the country in which the home is found.

2. Punch two holes in the top of each card. Loop a piece of string through each card and knot it to create a picture hanger.

3. Scatter pushpins over a bulletin board, and hang the cards on them. Invite children to approach the bulletin board display and examine the various pictures up close. Help them focus on the different attributes that make each house distinct by asking questions such as:

◎ What is this house made of?

◎ Does this house look big or small? How can you tell?

◎ Where does it look like this house is? By the water? In a forest?

4. Ask children to rearrange the pictures in different ways, for example grouping homes by shape, size, materials they're made of, geographic area, and so on. Create labels for the categories and tack them over the appropriate grouping.

**Note:** *Be sure children understand that homes are different not just across countries, but within them. While a family in a rural area in Ghana may live in a mud hut, one living in an urban area may live in a concrete-and-steel apartment house.)*

**ACTIVITY Extension** Reinforce theme vocabulary by creating a class dictionary of homes. Label pages of a pad of chart paper with each letter of the alphabet (one per page). As theme-related vocabulary comes up in activities and discussions, have children add the words to the appropriate pages and illustrate them.

*Houses and Homes* by Ann Morris (Morrow, 1995) offers a photographic look at dwelling places the world over. Reinforce geography skills by helping children find the countries mentioned in the book on a globe or wall map. Encourage children to think and talk about how an area's climate and available materials, as well as the lifestyle of a home's inhabitants, influence the type of structure built.

## Learning Center Link

*Place the picture cards of homes at the learning center and use them for sorting, comparing, writing, or speaking activities. Suggestions include:*

◎ *Find pairs of opposites among the pictures (big house/little house; house on land/house on water; round house/square house; and so on).*

◎ *Pick any two pictures at random, then brainstorm a list of ways in which the homes are similar and different.*

◎ *Compare one of the homes pictured with your own home. How are they similar and different?*

◎ *Which home in the display would you most like to live in? Explain why you'd like to live in that home, either orally or in writing.*

# Make It with Mud

**Students build houses with mud bricks, exploring the usefulness of a building material that has been around since ancient times.**

## Materials (for each group)

◎ old newspaper

◎ dirt*  (about 4 cups)

◎ mixing bowl or large plastic container

◎ measuring cup

◎ water

◎ several empty ice cube trays

◎ butter knife

* Not all soil is ideal for brick building—it needs a high clay content. Check to see whether the soil in your area will make sturdy bricks by mixing a small amount with water to make a soft mud. If you can form a ball that holds its shape, you can use it. If the ball crumbles, you can use air-drying clay for this activity (available from a craft or pottery store).

## Teaching the Lesson

**1** You may want to have children work in cooperative groups for this activity. Provide each group with the materials listed above, then lead them through the remaining steps.

**2** Cover a work surface with newspaper. Put the dirt in the bowl. Add 1/2 cup water and stir.

**3** Keep adding water a little bit at a time until you can mold the mud into a ball that holds its shape.

**4** Press the mud into the ice cube trays. Set aside any leftover mud to use later.

**5** Put the ice cube trays in a warm, sunny place. Let dry for 8 to 10 days.

**6** Again, cover a work surface with newspaper. Use the butter knife to pry the "bricks" out of the ice cube trays.

**7** Use the bricks to build a house. (Encourage children to be creative with their designs, stacking and arranging the bricks in different ways until they come up with a design they like. Once they've settled on a design, children can mix the leftover dirt with a little water to use as mortar to hold the bricks together.)

**8** Wrap up the activity by asking children what regions they think mud and clay homes would be best suited for. (Thick mud walls help keep hot air out and cool air in.) Are there places where mud homes might not be a good idea? What about in rainy areas? Expand the discussion to include other building materials as well, such as straw, wood, palm leaves, and cloth (for tents).

**ACTIVITY Extension** Display building materials at your learning center, including straw and sticks. Invite children to construct simple shelters—for example, tying bundles of straw together with string, then stacking them. They can use these structures, along with their mud homes, as props in dramatizing the story of *The Three Little Pigs*.

**Literature Connection** In *The Village of Round and Square Houses* by Ann Grifalconi (Little, Brown, 1986), a young girl from Cameroon describes the round clay homes and square brick homes of her community. Children might like to try using a fresh supply of mud to make the structures depicted in the book. (An overturned paper cup can provide the base for making the round mud homes; provide straw to use as roof thatching.)

# The Teeny Tiny House

**In this two-part activity, children make a shoe-box theater and perform a finger puppet play based on a traditional Russian Jewish folk tale about appreciating home—whatever the size.**

## Materials

- play script (see pages 25–26)
- shoe box
- scissors
- packing tape or other strong tape
- stick puppet patterns (see page 24)
- straws or craft sticks
- assorted art materials
- glue
- old decorating magazines (optional)

## Teaching the Lesson

**1** Divide children into groups of four to eight. Provide each group with a shoe box and the other art materials listed above. Use the illustrations (see page 23) to guide children in preparing their shoe box theaters.

**2** Give each group a copy of the puppet patterns. Have children color the puppets, then cut them out and tape each to a craft stick or straw. Using yarn, fabric scraps, and other supplies, children can add hair and other finishing touches to their puppets.

**3** Introduce the play to the children. Ask: Do you ever feel like your home is too crowded? Explain that they are going to hear and perform a play about a Russian family, unhappy about the small size of its house. Give each child a copy of the script. Read the play aloud as children follow along. Re-read the play several times to let children internalize the simple, repetitive plot. Invite children to join in as they feel comfortable.

4 Ask children to break into groups again. Depending on the size of the groups, have each member take on one or more roles in the play.

5 Have groups rehearse the play then perform it using the puppets and shoe box theaters.

6 Reflect on the folk tale's message: Did the size of the family's home change? What made the family think the home had become big? Do you think the size of a home is important? What do you think makes a house a home?

**ACTIVITY Extension** Talk about what we need in a home versus what we want a home to have. What would the children's ideal homes be like? Have them draw pictures and share their dream homes in a display.

**Literature Connection** A snail in *The Biggest House in the World* by Leo Lionni (Pantheon, 1968) learns the hard way that bigger isn't always better. Children may enjoy making tissue paper collages that depict the snail's brightly colored and elaborately designed—and, ultimately, cumbersome—dream house.

 **Learning Center Link**

*Make an audio recording of the play to share at your center. Display it along with a copy of the script. Children can listen to the tape as they follow along.*

## Directions for Making a Shoebox Theater

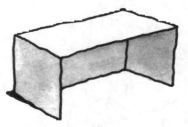

*Cut away one long side of the box.*

*Cut the sides of the strip you cut off to make a roof. Tape in place.*

*Decorate inside and out! Set the shoe box on the edge of a table so that the back hangs about two inches over the edge. Use strong tape to secure the box. Practice, then perform!*

# Puppet Patterns

woman

man

rabbi

cow

sheep

chickens

children

# The Teeny Tiny House

A play based on a Russian Jewish folk tale

**Characters:**

| | |
|---|---|
| storyteller | rabbi |
| man | chickens |
| woman | sheep |
| children (boy and girl) | cow |

**Storyteller:** Once there was a family who lived in a teeny tiny house. The house was so tiny that everyone was always bumping into one another. What an unhappy family!

**Man:** This house is so small and so crowded! What can we do?

**Woman:** Husband, you must go to the rabbi and ask him for advice. He will tell you what to do.

**Storyteller:** The very next day, the man went to see the rabbi.

**Man:** Rabbi, you must help my family. We live in a teeny tiny house. It is hard to move about, it is so small!

**Rabbi:** I know just what to do. You have chickens, yes? You must bring them to live with you in your house.

**Storyteller:** The man did not see how this would help, but he did what the rabbi told him. He went home and brought his chickens to live in the house.

**Chickens:** Buck-buck! Buck-buck! Buck-buck!

**Children:** Father, the chickens have laid eggs in our shoes!

**Woman:** Our tiny house is even more crowded than before. This won't do at all. Husband, you must go back to the rabbi.

**Storyteller:** The next day, the man paid another visit to the rabbi.

**Man:** Rabbi, we are still miserable. With the chickens inside, the house feels smaller than ever.

**Rabbi:** I know just what to do. You have a sheep, yes? Bring it to live with you in your house.

**Storyteller:** Again, the man was puzzled. But he followed the rabbi's advice.

**Sheep:** Baaaahhhh! Baaaahhhhh! Baaaaahhhh! Baaaahhhhhh!

**Chickens:** Buck-buck! Buck-buck! Buck-buck!

**Children:** Father, the sheep has crawled into our bed. We have nowhere to sleep.

**Woman:** How awful! I am forever tripping over the chickens and this sheep. Husband, tell the rabbi we cannot live this way.

**Storyteller:** The next day, the man set off to see the rabbi again.

**Man:** We are more miserable than ever. The chickens leave feathers everywhere and the sheep chews things up. You must help us!

**Rabbi:** I know just what to do. You have a cow, yes? Bring it to live with you in your house.

**Storyteller:** And that's just what the man did.

**Children:** Oh, no! Not the cow, too!

**Cow:** Moooooo! Mooooooo! Moooooo! Mooooooo!

**Sheep:** Baaahhhh! Baaaahhhh! Baaaahhhh! Baaahhhhh!

**Chickens:** Buck-buck! Buck-buck! Buck-buck!

**Woman:** Enough is enough! There is no room to move. Husband, go back to the rabbi now!

**Storyteller:** The man went to see the rabbi right away.

**Man:** Rabbi, how unhappy we are! The cow takes up the whole kitchen.

And the noise in the house! It is terrible! What can we do?

**Rabbi:** I know just what to do. Bring your chickens, your sheep, and your cow back out to the barn.

**Storyteller:** The man rushed home. He did just what the rabbi told him. Out went the cow.

**Cow:** Moooooo! MOOOOO!

**Storyteller:** Out went the sheep.

**Sheep:** Baaahhhh! BAAAAAHHHH!

**Storyteller:** Out went the chickens.

**Chickens:** Buck-buck. Buck-BUCK!

**Man:** Ah! How roomy it seems now!

**Children:** So much space to play and run about!

**Woman:** How lucky we are to have such a wonderful house!

**Storyteller:** And the man, the woman, and their children never complained about their home again. Without growing even an inch in size, their teeny tiny house had become to them the roomiest home in the world!

**The End**

# Animals at Home

**A**nimals need homes for the same reasons people do: to protect themselves from the elements and enemies, and to have a place to rest, store food, and raise their young. In this section, children discover animals at home in an oak tree, watch a beaver building its lodge step-by-step, and peek inside the towering mud home of the tiny termite.

## BACKGROUND NOTES

Animals may find a home—such as a tree trunk or cave—or build one. Animal architects use materials available in their environment (grass, leaves, twigs, and so on), or materials they manufacture in their own bodies (wax, saliva, silk, and so on). Homes animals build vary widely in complexity from simple holes in the ground to intricate structures with rooms, tunnels—even cooling systems!

# Anybody Home?

**Children begin their investigation of animal homes by constructing simple lift-the-flap pages that show animals living in a tree.**

## Materials

- ◎ Anybody Home? tree pattern (see page 32)
- ◎ construction paper cut into 3-by-3-inch squares or self-sticking notes
- ◎ scissors
- ◎ tape
- ◎ chart paper

## Teaching the Lesson

1. Tap children's prior knowledge about animal homes. Have they ever seen a place where an animal lives? Where was it? What did it look like? Who lived in it?

2. Ask children what a tree has in common with an apartment building. Write their ideas on the chalkboard. Encourage children to recognize that both apartment buildings and trees provide lots of homes for living creatures. Chart some of the animals that make their homes in trees, for example birds, squirrels, wasps, various insects, raccoons, and moles (in the roots).

3. Have children make lift-the-flap tree homes for animals. It will be helpful to make and display a model first.

- ◎ Give children copies of a tree pattern. Have them draw pictures of animals living in the trees.
- ◎ Have children place the paper squares over their animals and tape the sides or top to make flaps.

- ◎ Have children write Who Lives Here? on their flaps. (Or let them write riddles about the hidden animals' identities.)

4. Put children's pages together with O-rings or yarn to make a riddle book of animals at home in a tree. Place the book at your learning center. As children look at the pages, challenge them to guess each animal's identity before lifting the flap.

5. Follow up by creating a class chart with three separate categories: Finders, Diggers, and Builders. Ask: Which animal or animals in the tree do you think found a home and moved right in? (*raccoon, insects*) Which do you think dug a hole in the ground for itself? (*mole*) Which do you think actually built something? (*robin, squirrel, wasp*) Record children's responses on the appropriate part of the chart. Challenge children to add more animals to the chart as you move through the unit.

**ACTIVITY Extension** Go on a house hunt. Take children to a nearby park to look for places animals make their homes.

**Literature Connection** *The Apartment House Tree* by Bette Killion (HarperCollins, 1989) and *Cactus Hotel* (Holt, 1991) by Brenda Guiberson take a look at the community of animals who make their homes in a woodland tree and a saguaro cactus, respectively. Have children add the new animals they meet in these books to your class chart of Finders, Diggers, and Builders.

**Learning Center Link**

*Provide additional copies of the tree pattern. Invite children to make new lift-the-flap trees that show more animal homes they discover.*

# Busy Beaver Builders

**Children make mini-books that describe step-by-step how beavers build their homes, and participate in pocket chart reading and sequencing activities.**

## Materials

- ◎ reproducible mini-book (see pages 33–34)
- ◎ sentence strips
- ◎ pocket chart

## Teaching the Lesson

1 Following is the text that appears in the mini-book students will make. Write each line on a sentence strip. Put the sequenced sentence strips in the pocket chart, and gather children around.

> Busy, busy beavers, lots to do.
>
> Find some trees, now chew, chew, chew.
>
> Move those logs along, quick, quick, quick.
>
> Pile them up high, stick by stick.
>
> Add some stones and leaves, pat, pat, pat.
>
> Cover them with mud, splat, splat, splat.
>
> Now dig a cozy room, deep inside the dome.
>
> Happy, happy beavers, now you have a home!

You may also want to make picture cards for the pocket chart. You can use the art from the mini-book to make the cards, which will provide students with visual clues to the text during reading activities.

2 Ask children if they have ever heard the expression *busy as a beaver*. What does it mean? How do they think beavers got this reputation? Tell children they're about to find out. Before reading the story, tap children's prior knowledge about beaver homes: Do they know what a beaver's home is called? (*a lodge*) Where do beavers build them? (Students may be interested to know that beavers' homes have underwater entrances. How do children think this would help protect beavers from enemies?)

3 Read the text on the pocket chart through once, emphasizing the rhythm and rhyme with your voice and pointing to the text as you read. During a second reading, ask children to point out words they hear repeated. (*chew, chew, chew; quick, quick, quick; and so on*) Invite children to join in the reading.

4 Remove the sentence strips from the chart. Give each strip to a child, then ask children to re-sequence the strips in the chart pockets.

5 Make double-sided photocopies of the mini-book pattern. (Be careful not to invert the text on the back of the page.) Guide children through these steps to make the book:

- ◎ Place the side of the page labeled with panels A and B faceup on the desk.
- ◎ Cut along the solid line to separate the panels.
- ◎ Place panel A on top of panel B.
- ◎ Fold the panels in along the dashed line to make the book.
- ◎ Staple along the left edge to hold the pages in place.
- ◎ Color pictures to go with the words.

6 Invite children to follow along in their mini-books as you read the text aloud again. Use the oral cloze technique to get children to focus in on individual words. Encourage children to use picture and contextual clues to decode the language.

7 To help boost children's confidence and sense of accomplishment as readers,

safety pin *Ask Me to Read Busy, Busy Beavers* buttons to their shirts or jackets at the end of the day. (See reproducible sample, below.) Be sure they take their mini-books home with them. The buttons will serve as prompts for at-home sharing.

**ACTIVITY Extension** Encourage children to invent hand motions and or body movements to accompany the action in the story (chewing trees, piling up sticks, patting mud, digging a room). Using the hand motions during a pocket chart reading will enhance children's comprehension of the story's events.

**Literature Connection** *Busy Beavers* by M. Barbara Brownell (National Geographic Society, 1988) is above-level for K-1 students, but it offers some stunning photographs of beavers busy building homes. Share some of the photos and information in the book. Then let children work together to

 **Learning Center Link**

*Copy the text of the book onto chart paper. Drop the last word off the second line in each couplet, leaving a blank space for it on the chart. Write these words on cards, then glue Velcro to the back of each card and to the blank spaces on the chart paper. Children can then try to add the rhyming words that complete each couplet.*

make a mural depicting the many animal homes in and around a beaver pond with the beaver lodge as a focal point.

## SCIENCE

# Life in a Castle

In this two-part activity, children take a look at a termite's home—inside and out—then investigate how a termite's tower of mud protects it from the scorching East African heat.

## Materials

◎ Life in a Castle reproducible (see pages 35–36)

FOR EACH GROUP:

◎ a large plastic tumbler (taller than 6 inches)

◎ air-drying clay

◎ butter knife

◎ 2 stand-up thermometers (no taller than 6 inches)

## Teaching the Lesson
## Part I: A Tower of Tunnels

1 Ask children: Have you ever seen an insect home? Do you think insects build their homes? Allow time for discussion, then explain that they are going to learn about one of the most amazing homes constructed by insects: termite mounds.

2 Make double-sided photocopies of pages 35 and 36, being careful not to invert the picture on the back of the page. Provide each child with a copy, and allow time for students to study the outside view of the termite mound. Explain that the tiny termites build the mounds out of soil (mixed

with their own saliva) and that some mounds are as tall as a giraffe!

3 Tell children that millions of termites may live in a single mound, working together as a community. They dig out rooms for resting, storing food, and raising babies. The rooms are connected by tunnels. The queen termite lays all the eggs. She has her own special room in the castle. Show children how to hold the page up to the light so that the inside of the mound becomes visible. Ask students to describe what they see. Can they spot the queen? (Hint: She's much bigger than all the other termites.) Let children estimate the number of termites inside the mound, then count. Can they help the termite at the top find its way to the queen?

## PART II: Beating the Heat

Explain that termite mounds are found in parts of East Africa, where it is extremely hot. Have students locate this region on a map then guide them in groups through the following experiment to discover how the termite mound helps protect its inhabitants from the heat.

1 Overturn the tall plastic tumbler and cover it with a thick layer of clay. Set it in a sunny spot.

2 Allow the clay to dry for a day or two. Before the clay becomes completely dry, pull out the cup. Use a butter knife to pry the cup loose, if necessary.

3 Stand the two thermometers side by side on a sunny window ledge. Make sure they both read zero degrees. Cover one of the thermometers with the clay mound. Leave the thermometers on the ledge for several hours.

4 Ask children to predict whether the temperature readings of the two thermometers will be the same or different. Help children read the temperatures on both thermometers after removing the mound. Students will notice that the temperature of

the thermometer covered by the clay mound is lower. Ask them to draw conclusions about why termites in hot regions build mud homes.

**ACTIVITY Extension** Tell children that termite mounds are the tallest animal homes, and that they may tower as high as 25 feet. In the hallway or another open area, enlist children's help to measure off this length. Then have children use various objects as non-standard units of measure for size comparisons. For example: How many chalk erasers high is a termite tower? Create a chart to illustrate these height comparisons.

**Literature Connection** Ant colonies, with their systems of tunnels and rooms and hierarchy of workers, are organized in much the same way as those of termites. Share the book *Anthony Ant's Treasure Hunt* by Lorna Philpot (Random House, 1996). Children will enjoy helping the title character find his way through a series of tunnel mazes. As a follow-up, set up an ant farm so that students can observe firsthand the at-home activities of these industrious insects. (See Professional Resources, page 7.)

### Learning Center Link

*Leave copies of the Life in a Castle reproducible in your learning center. Children can try to find their way through the maze of tunnels on the back of the page, and pick up fast facts about termites along the way.*

# Anybody Home?

Add some stones and leaves,
pat, pat, pat.

5

Find some trees, now chew,
chew, chew.

2

A  B

Now dig a cozy room, deep
inside the dome.
Happy, happy beavers, now you
have a home!

7

# Busy,
# Busy Beavers

Move those logs along, quick, quick, quick.

3

Pile them up high, stick by stick.

4

Busy, busy beavers, lots to do.

1

Cover them with mud, splat, splat, splat.

6

# Life in a Castle

Termites live in tall towers made of mud, like this one. Hold the page up to the light.
What do you see?

# Life in a Castle

Help the termite at the top of the tower bring food to the queen.
Use a pencil or crayon to trace your path.

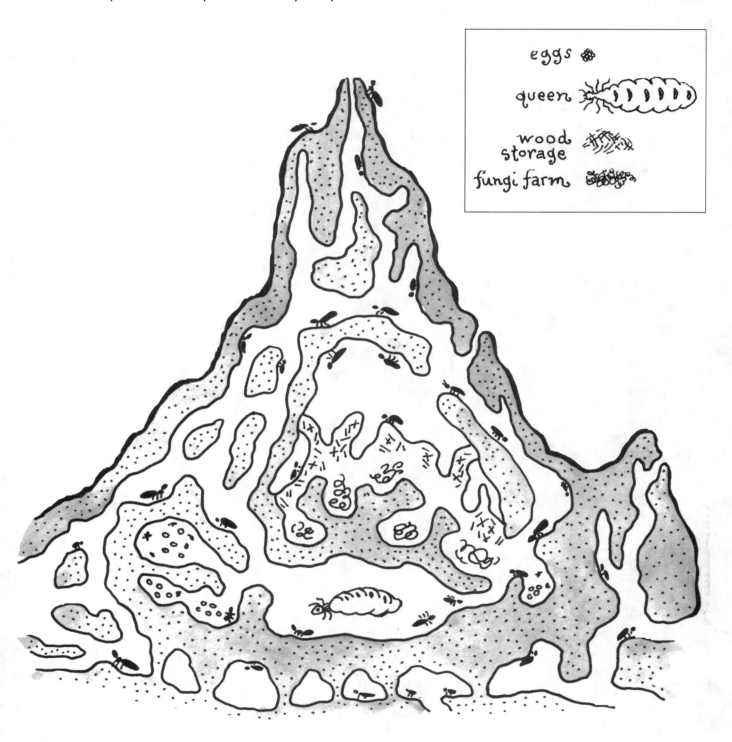

# NOT JUST FOR BIRDS:
# A Look at Nests

**H**ave your students ever seen a tidy mud-and-straw nest fashioned by a robin? A robin's neat nest is just one of the many marvelous structures crafted by birds. However, birds aren't the only nest builders. Alligators, mice, squirrels, wasps, ants—even some kinds of fish— all build nests. In this section, children will investigate the nests of a bevy of birds, as well as nests made by other animals, such as the European harvest mouse.

## BACKGROUND NOTES

Nests fulfill a very specific purpose: they provide a soft, warm, safe place for animals to lay eggs or give birth to live young. They also serve as a nursery in which to raise offspring until they are self-sufficient. Animals grab whatever materials they can get their hands—or beaks or feet—on: stones, twigs, grass, mud, feathers. Some birds even pluck the wool right off a sheep's back! The craftsmanship of nest-builders varies considerably. For example, the Antarctic Adélie penguin simply piles up pebbles to make its nest, while African weaverbirds wind grass into elaborate basketlike structures.

# Nest Builders

**In this activity, children give birds a hand at building nests and discover some materials that birds like best.**

## Materials

- ◎ chart paper
- ◎ paper plates (two per group)
- ◎ stapler
- ◎ scissors
- ◎ assorted yarns, string, dryer lint, bits of cloth, small twigs, grasses, and so on
- ◎ hole punch

## Teaching the Lesson

1. Ask children if they have ever seen a bird's nest. What did it look like? Why do they think birds build nests? Why do they think birds need a special place to lay eggs? (*to keep them safe from enemies, to keep them from cracking, and to keep them warm*) Explain that different birds build different kinds of nests using different materials to make nests of different shapes and sizes. Ask students what they think might make good nest-building materials.

2. Share these nest descriptions with children. Follow up by having children list materials these and other birds use.

- ◎ The **tailorbird** makes its nest from a leaf. It uses silklike thread to sew two sides of the leaf together. This makes a neat pocket where the bird can lay its eggs.

- ◎ **Weaverbirds'** nests look like hanging baskets. Weaverbirds use their beaks and feet to weave green grass and vines into a ball shape. The door is at the bottom.

- ◎ **Cliff swallows** make their homes on rocky cliffs. They grab clumps of mud with their beaks, then stack up the mudballs to make round nests with roofs.

3. Have students work in groups to make nest-builder boards for birds in your area. Directions follow.

- ◎ Glue together two sturdy paper plates.

- ◎ Cut slits around the edge of the plates. (You may wish to precut the slits for students.)

- ◎ Pull pieces of string, thread, and yarn (untwist the strands first), and other materials through the slits, letting the ends hang down. Glue some dryer lint to the center of the plates.

- ◎ Punch a hole in the edge of the plate and loop a piece of string through it. Tie the nest-building boards to tree branches. Watch to see which materials the birds take.

**ACTIVITY Extension** Most birds (eagles and storks are exceptions) abandon their nests once their young learn to fly; they build new nests the following spring. Look for old birds' nests in the fall or winter and display them at the learning center. If possible, provide a field guide to identify the bird who built the nest. Children, wearing plastic gloves, can pull apart the nest and investigate the materials used to make it, as well as its construction. If you can find several old nests to examine, create a chart comparing the design and building materials in each.

╔═══════════╗
║ Literature ║ In P.D. Eastman's classic *The*
║ Connection ║ *Best Nest* (Random House,
╚═══════════╝ 1968), the search for the
perfect nest leads Mr. and Mrs. Bird back to their own home. Make a list of all the materials that Mr. and Mrs. Bird gather to build their nest. How do the materials they used compare to the ones in students' nest-builder boards?

## Learning Center Link

*Provide additional materials (paper plates, stapler, yarn, straw, string, dryer lint, and so on) for making nest-builder boards. Children who are interested can make their own to bring home and hang on a tree.*

MATH
∘ ∘ ∘ ∘ ∘ ∘ ∘ ∘ ∘ ∘ ∘ ∘ ∘ ∘ ∘ ∘ ∘ ∘ ∘ ∘ ∘ ∘

# Big, Bigger, Biggest

**Children use picture cards to compare bird nest sizes.**

## Materials

◎ Big, Bigger, Biggest reproducible (see page 43)

◎ scissors

◎ chart paper

◎ whole walnut in shell, teacup, cereal bowl (optional)

## Teaching the Lesson

1. The reproducible activity page offers a size sampling of different birds' nests. Photocopy the page and distribute it to the children.

2. Explain that some birds build very big nests and some birds build very small nests. Together, look at the activity page. Show children how to look at the object pictured beneath each nest to figure out the nest's size. For example, a walnut appears underneath the picture of the hummingbird's nest. This means that the hummingbird's nest is about the size of a walnut—pretty small!

3. Write the words *small, smaller, smallest* and *big, bigger, biggest* on the chalkboard or chart paper. Review with children the way in which these words are used to compare sizes.

4. Ask children to cut apart the three nest cards on the top part of the page and set aside the bottom of the page to use later. Have children put the nests in size order, following the pattern small, smaller, smallest. Children can paste the size-ordered nests on another paper, then label them: *small, smaller,* and *smallest.* (You may want to have on hand a walnut, a teacup, and a cereal bowl to help students visualize the actual nest sizes.)

5. Repeat this process using the picture cards on the bottom of the page, but this time have children follow the size pattern big, bigger, biggest, labeling the nests accordingly.

 Make edible, size-accurate models of hummingbird nests by following these directions:

◎ In a large bowl mix together 1 1/2 cups sunflower seeds, 1 1/2 cups raisins, 1 1/2 cups granola, 1 cup honey, and 1 cup peanut butter.

◎ Put 2 cups shredded, toasted coconut in another large bowl. (To toast coconut, spread on a cookie sheet and bake at 350°F till golden. Cool.)

◎ Spoon a walnut-sized chunk of the mixture into each child's hands. Have children roll the mixture into balls, then roll in coconut to coat.

◎ Each child can press his or her thumb into the center of the ball to create a deep impression. Drop two small jellybeans or yogurt-covered raisins (which approximate the size of hummingbird eggs) into each nest as a finishing touch.

**Literature Connection** Wrap up by asking students to account for the differences in birds' nest sizes (*size of birds, number of birds living together, number of eggs laid*). Share *Animal Architects* (National Geographic Society, 1987) for photographs and detailed descriptions of birds' nests, including some of those pictured on the reproducible.

 **Learning Center Link**

*Different birds lay different numbers of eggs in their nests. This is known as a bird's clutch size. Write the following clutch sizes on chart paper, and post it at your learning center.*

> **eagle:** 2 eggs
> **robin:** 4 eggs
> **blue jay:** 7 eggs
> **duck:** 8 eggs
> **ostrich:** 10 eggs
> **chickadee:** 15 eggs

**Note:** Clutch sizes can vary within a species. The numbers shown are approximate; they indicate the upper range of each bird's clutch size.

*Cut the top off a half-dozen–sized egg carton. Write the name of each bird on a small piece of paper. Tape it to a toothpick, and stick the toothpick into a small bit of clay. Place a toothpick inside each egg compartment. Set the carton and a jar of dried beans, or other counters, near the egg clutch chart. Invite children to count out the number of eggs in each bird's clutch, placing the counters in the appropriate compartments. Partners can check each other's work.*

# Hide-Away Home

**Share a poem about a mouse's house, then pick out rime patterns for a phonics and spelling lesson on the long *e* sound.**

## Materials

- pullout poster
- chart paper
- self-sticking notes

## Teaching the Lesson

1. Ask children if they know of any animals besides birds that build nests. Display the poster and introduce them to the harvest mouse. Provide children with a little background information about this tiny creature and its cozy home:

   - A harvest mouse is only 2.5 inches long, and it weighs one third of an ounce. (Try to find size comparisons.)

   - It makes its home in wheat and oat fields (hence its name).

   - It builds its home high in the grasses, away from hungry snakes.

   - It uses grasses to weave a neat, hollow, ball-shaped nest that is about the size of a tennis ball.

   - It lines the inside of its nest with thistledown and cattail fluff to make a soft, warm place to give birth to three to five babies.

2. Read the poem on the poster aloud, emphasizing the rhythm and rhyme. Point to the text as you read each word. Invite children to join in on a re-reading.

3. Discuss with children what parts of the poem they like. What do they think the poet has done well? Ask children specifically about the sound of the poem. What makes it sound musical? Ask children to point out the rhyming words they hear, both at the end of lines and within them. Tell children that there is something else that makes the poem musical—the long *e* sound is repeated over and over again.

4. Read the poem aloud again, emphasizing the words wee, see, green, neat, and sweet. Ask children to raise their hands every time they hear the long e sound. Use self-sticking notes to flag words with the long *e* sound.

5. Draw students' attention to the two different spelling patterns used to achieve the long *e* sound in this poem: *ea* and *ee*. Focus children's attention on the word *chunks*, or *rimes*: *eet* and *eat*. (The rime is the part of the syllable that follows the initial consonant sound; it includes the vowel and any consonant(s) that follow it.)

6. As a class, brainstorm a list of words that contain the *eet/eat* sound in them. Invite children to offer their ideas about how to spell the words suggested, providing assistance as necessary.

7. Write each rime—*eet* and *eat*—at the top of a two-column chart. Ask children to look at the words they generated. Invite them to write each word in the appropriate column, based on its spelling pattern.

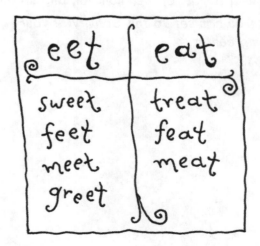

| eet | eat |
|-----|-----|
| sweet | treat |
| feet | feat |
| meet | meat |
| greet | |

**8** Display the chart in your classroom. Children can refer to the list to help them decode the sound of words with similar spelling/rime patterns in other reading activities. They can also use the list as a reference tool for their own writing.

**ACTIVITY Extension** Use the poem as a springboard for a language arts lesson on synonyms and antonyms. Ask students to find words in the poem that mean small (*wee, little*). Explain that these words are *synonyms*; they have similar meanings. Next, ask them to find a word that means the opposite of small (*big*). Explain that words that are opposite in meaning, such as *little* and *big*, are called *antonyms*. Brainstorm a list of additional synonyms for *little* and *big*. Write these words on self-sticking notes. Invite children to stick the notes on the poster to replace the words *wee, little*, and *big* in the poem during subsequent re-readings.

**Literature Connection** *And So They Build* by Bert Kitchen (Candlewick Press, 1993) includes information about nests fashioned by other animals, including fish, frogs, and wasps. Share some of the text and illustrations. Then provide students with stacks of nature magazines such as *My Big Backyard, Owl*, and *Ranger Rick*. Challenge children to find pictures of other animal nests, then sort the pictures according to the type of animal that built each nest (bird, reptile, mammal, and so on).

**Note:** *You can set up this activity at your learning center.*

## robin

size of nest

↓

teacup

## blue jay

size of nest

↓

cereal bowl

## hummingbird

size of nest

↓

walnut

## eagle

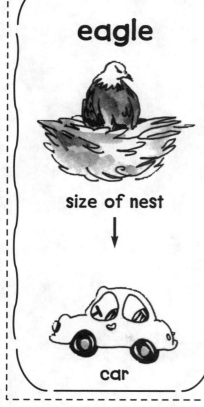

size of nest

↓

car

## stork

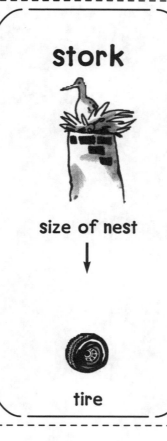

size of nest

↓

tire

## sociable weaver

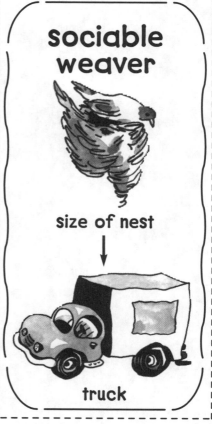

size of nest

↓

truck

# A Celebration of Home

Be it a humble cottage, a grand mansion, or a nest in a tree, homes fulfill a universal need for shelter. In this section, students will take one last look at the breathtaking variety of homes inhabited by both people and animals, then draw conclusions about their shared characteristics as well as their distinctive qualities. A culminating project will provide your class with an opportunity to demonstrate and celebrate what they've learned about homes during this thematic unit.

# Open House

In this two-part project, children form construction companies to build models of various human and animal homes. They then share their projects with one another and with family during an open house party.

## Materials

- assorted boxes (shoe boxes, cereal boxes, spaghetti boxes, gift boxes, and so on)
- clean juice and milk containers
- natural materials (mud, twigs, grass, straw, leaves, pebbles, and so on)
- building materials (blocks, Legos, sugar cubes, craft sticks, clay, and so on)
- miscellaneous junk materials (paper towel tubes, soda straws, toothpicks, fabric, coffee cans, oatmeal containers, plastic flowerpots, old lamp shades—anything that might be useful for building)
- art supplies (tape, glue, scissors, colored paper, mural paper, paint and paintbrushes, crayons, markers)
- real estate section of a local newspaper (optional)

**Note:** *You may want to send home notes to parents requesting materials and/or go on a nature walk with children to gather materials from the outdoors, being careful not to disturb any animal communities and taking only what you need.*

## Teaching the Lesson

### Part I: Building the Homes

1. Organize materials at work tables. Display posters and pictures of all kinds of homes—human and animal—around the room. Be sure your reading corner is well-supplied with books on the topic.

2. Divide children into cooperative groups. Explain that each group is going to form a construction company to design or build specific types of homes. Have each group choose a name/category.

3. Allow time for group members to brainstorm and research ideas about types of shelters they will make, and how to approach the project. Will they collaborate on one or two types of homes, or will each group

## Construction-Company Starters

**Children can choose from these names or come up with their own.**

- *Small World Builders (homes around the world)*
- *Animal Architects, Inc. (animal homes, other than nests)*
- *Nests R Us (bird and other animal nests)*
- *The Tradition Commission (traditional Native American and Native Alaskan homes, such as igloos, hogans, tepees, and pueblos)*
- *Just U.S., Inc. (famous U.S. homes, such as the White House, Lincoln's log cabin, Mt. Vernon, and so on; homes of historical significance in your community)*
- *The Dream Team, Inc. (futuristic homes)*
- *Fairyland Builders (well-known houses from literature, such as Hansel and Gretel's gingerbread house, the three little pigs' homes, the oversized shoe-house of "The Old Woman Who Lived in a Shoe," distinctive homes from children's books, such as Baba Yaga's round hut that stands on chicken legs, from the popular Patricia Polacco book by the same name)*

member make a separate home? Would they like to construct a model, paint a mural, design a diorama, or create something using a computer drawing program?

4 Invite group members to examine the available materials at the work table. What can they use to make their designs or models? Circulate among the groups to listen to ideas and offer input. If children are struggling, you might want to hold a class brainstorming session to spur creative thinking. Some suggestions to share:

◎ Use sugar cubes and glue to make an igloo.

◎ Pile various size gift boxes on top of one another to make a pueblo.

◎ Use soda straws and fabric to make a yurt.

◎ Stand a milk carton on toothpicks to make a coastal Thai home.

◎ Overturn a small, square gift box bottom on a shoe box top to provide the framework for a Dutch houseboat on a converted barge.

◎ Cover an overturned plastic flowerpot with mud to make a flamingo nest.

◎ Spread papier-maché pulp over a balloon for a neat wasp's nest.

◎ Stick graham crackers together with peanut butter and festoon them with yogurt-covered raisins to replicate the edible home in Hansel and Gretel.

**ACTIVITY Extension** Bring the real estate section of a local paper to class. Discuss its purpose, then read some of the ads aloud. Children can then write or dictate ads for the homes they made. Encourage children to highlight the best features of their homes in their sales pitches. Publish a class real estate section and display it on the day of your party.

# Part II: Planning the Party

1 Pick a day on which to hold an Open House party to display students' construction projects and other theme work. Invite parents/caregivers so that children can share their accomplishments with others. Have children design lift-the-flap invitations for a special treat. Here's how.

◎ Draw a simple house with several windows and a door on a piece of copy paper and reproduce. (See sample, below.) On each copy, cut flaps where the windows and door are. Spread paste along the outer edges of the back of the paper. Press firmly to another piece of paper. Lift the flaps and write party information inside.

2 Do some advance party planning, encouraging students' input as you decide on decorations, refreshments, and activities. Following are suggestions for incorporating some of the activities and ideas in this book into your celebration:

◎ Make and serve hummingbird nest treats as snacks. (See page 39.)

◎ Plan performances of the finger puppet play "The Teeny Tiny House." (See pages 25–26.)

◎ Invite parents to participate in read-alouds that celebrate home, such as "From My Room" by Wendy Murray (see page 16) and "The House of the Mouse" by Lucy Sprague Mitchell (poster), or a favorite story or poem of their own.

◎ Display students' work from the unit for your guests to view.

**3** Set up tables to display each group's construction projects. Position tables around your room so that students and guests can easily circulate among the displays. Groups may want to create a banner bearing their construction company's name to hang over their respective tables.

**4** On the day of the party, encourage groups and guests to circulate around the room. Group members can take turns stationing themselves near their projects and offering their services as tour guides. Tour guides shoud be ready to talk about their group's work, share facts and information, and answer questions.

**ACTIVITY Extension** Following the party, review and assess what students have learned about homes with a quick charting activity. Draw the outline of a house on a piece of chart paper. Divide the house into two columns. Label one column Same, the other Different. Ask children to look around at all the homes *they have built or designed (animal homes, people homes, fantasy homes, and so on).*

Ask: What do all of these homes have in common? List students' ideas on the left side of the chart (for example, they all provide protection from weather, protection from enemies, a place to rest, raise young, store food, and so on). In the right-hand column, children can list the ways in which the homes are different (for example, in size, shape, design, building materials, occupants, location, and so on).

**Literature Connection** Two literature selections just right for sharing at your open house party are the exuberant poem "Home! You're Where It's Warm Inside" by Jack Prelutsky (from *The Random House Book of Poetry for Children*, Random House, 1985) and the lovely and lyrical picture book *I Know a Place* by Karen Ackerman (Houghton Mifflin, 1992). Before the party, invite children to use both as models for writing poems or paragraphs about what makes their own home special before your open house. Children can share their writings at the celebration.